I0470310

"I've worked with Cheryl at Mobile Apps For Small Business for a couple years now and I know first hand she knows what she's talking about. I highly recommend working with her. You are going to quickly see tangible results adding dollars to the bottom line. One of the best things about working with Cheryl is she operates with a high level of integrity and runs her business that way."

James Cusick
Houston, Texas

"To say the process has been smooth and all around great would be an understatement. She has been fantastic. If you need any kind of App work I highly recommend Cheryl. She is awesome. We had ours done in a week. I couldn't believe it. Bottom line, she made everything really easy. I 100% recommend Cheryl. You cannot go wrong. She is the best out there."

Danny Ives
www.annpolismma.com

"I've had the pleasure of working with Cheryl Paulsen with Mobile Apps For Small Business and I can tell you the process has been amazing, the customer service has been top notch. I highly recommend her, you are going to get treated very well and you better work with her before your competition does!"

Rafael Lovato
www.okbjj.com

"*The Book on Mobile Apps* lives up to its name. It truly IS 'The Ultimate Guide for Small Businesses'. Cheryl Paulsen delivers creative thinking, proven strategies, and vast knowledge about the value of mobile apps for small businesses. Her book is timely, informative, and a must-read for any entrepreneur today."

—Raymond Aaron,
Author of *Double Your Income Doing What You Love*

CLAIM YOUR FREE GIFTS NOW!

MORE BONUSES AVAILBLE AT
www.TheBookOnMobileApps.com

BECOME A SMALL BUSINESS
RENEGADE™—FOR FREE!

($407.97 Worth of Pure Money
Making Information)

Cheryl Paulsen is offering an incredible opportunity for you to see _why_ Small Business Renegades™ is _**the place**_ for small businesses that are seeking to navigate the treacherous waters of successfully marketing your business. Get $407.97 worth of pure _**money-making**_ information—including a _free_ month as an 'Elite' Gold Inner Circle Member of Small Business Renegades™.

Here's what's included with this free offer:

- **Small Business Renegade University: Webinars** (Value: $387.00)

  Learn how to change the game and fuel the app addiction so you can PUSH profits to put more $$$ in your pocket.

  - How any business can multiply income by 10X

  - The importance of a mobile app in your sales process

  - How to use the latest technology to instantly reach your clients

  - 3 most important steps to monetizing your mobile app

- **'Elite' Gold Inner Circle Membership** (Value: $19.97)

  An issue of the Small Business Renegade Marketing Letter. Each issue overflows with the latest marketing and mobile strategies, plus terrific examples of "What's Working Now" strategies, timely marketing news, trends, and ongoing teachings. As soon as it arrives, you'll want to find a quiet place, grab a highlighter, and devour every word!

- **Exclusive "Members Only" Perks:**
  - o Special member teleconferences every month
  - o Restricted access website
  - o Continually updated *Million Dollar Resource Directory* with valuable contacts and resources used by Cheryl and her clients.

To activate your ***most incredible free gift ever***, you pay only a one-time charge of $3.95 ($6.95 for international subscribers) to cover shipping.

After your one-month free test drive, you will automatically continue at the lowest price of $19.95 per month. Your credit card will not be charged for the monthly fee until the beginning of the month after your free trial. You get a full month to test and profit from all the powerful techniques and strategies that will change your mobile game. It's impossible for you to lose, because if you don't absolutely *love* everything you get, just cancel your membership before the next month and never get billed a single penny!

You can cancel your membership at any time by calling Mobile Apps for Small Business at 410-827-6236 or faxing a cancellation note to 888-378-1118 (Monday-Friday, 9 am – 5 pm).

Name _____

Business name _____

Address _____

City _____ State _____

Postal code _____ Country _____

Email* _____

Phone _____ Fax _____

Credit Card Instructions to cover $3.95 ($6.95 Int'l) for shipping:

_____ Visa _____ Master Card

_____ American Express _____ Discover

Name on card

Credit card number _____ Exp date _____

Signature _____ Date _____

*EMAIL IS REQUIRED TO NOTIFY YOU ABOUT THE SMALL BUSINESS RENEGADES WEBINARS

Providing this information constitutes your permission for Mobile Apps For Small Business to contact you regarding related information via, mail, email, fax, and phone.

FAX BACK TO 888-378-1118
Or mail to: 600-B Abruzzi Drive #125, Chester, MD 21619

PUSHing Profits

The Book on Mobile Apps

Cheryl Paulsen

BALBOA.
PRESS

A DIVISION OF HAY HOUSE

Balboa Press books may be ordered through booksellers or by contacting:

Balboa Press
A Division of Hay House
1663 Liberty Drive
Bloomington, IN 47403
www.balboapress.com
1-(877) 407-4847

ISBN: 978-1-4525-5373-3 (sc)
ISBN: 978-1-4525-5372-6 (e)

Printed in the United States of America

Balboa Press rev. date: 06/25/2012

TABLE OF CONTENTS

Foreword. ix
Introduction. xiii
Chapter 1:   Past, Present, and Future of Mobile 1
Chapter 2:   7 Reasons Your Business Is Craving a
Mobile App 8
Chapter 3:   So You Think _Your_ Business Is Different?!. 16
Chapter 4:   5 Mistakes Small Business Owners
Make Building An App. 23
Chapter 5:   9 Steps to App Success 30
Chapter 6:   Tips for Apple Compliance 39
Chapter 7:   Marketing Keys to Success 45
Chapter 8:   Post-App Hangover: 3 Steps to
Monetize Your App .53
Appendix A: Technophobia? Not Any More. 61
About the Author. 67

FOREWORD

I'm a firm believer in making your business work harder for you than you work for it. As Michael Gerber, author of the *E-Myth* book series says, too many entrepreneurs spend valuable time working **_in_** their business when they should be working **_on_** their business.

When you're invested in making your business a success, you have to broaden your vision and see opportunities that might not be glaring you in the face.

I was broke at the age of 39. Two years later, I was a millionaire. I did it by seeing past what people told me was impossible. I ignored the naysayers. Sometimes, you have to allow yourself to dream bigger than you ever imagined.

Steve Jobs and Apple did that. As a result, the way we listen to and buy music has been revolutionized. The cell phone evolved into a higher level of intelligence with the launch of the first iPhone in 2007, followed by the brilliance of the iPad in 2010.

What these devices have done is create opportunity for the smart marketer. Companies have emerged to feed the unending appetite for accessories for their mobile devices. And software developers have discovered that the consumer's desire for mobile applications is equally insatiable.

So, the question is, are you going to leverage a fantastic opportunity to take your brand to new heights by developing a mobile app? Or are you going to sit by and watch your competitors beat you up and out?

With consumers being not just influenced by mobile media but reliant on it, business owners have a prime chance to reach out and grab their audience wherever they are. When have you ever had that power?

I'm excited by the new wave of mobile apps and the developers who continue to raise the bar on the technology. As someone who has made a career out of helping small businesses brand themselves more effectively, I'm thrilled that this power tool is now within reach of anyone—not just the corporate giants with the deep IT and marketing budgets. It's doable today for even the local shop owner to have an app on Apple's App Store and the Android Marketplace—next to the likes of Amazon, Facebook, and other global brands.

It's not only doable—but easy. And that's where Cheryl Paulsen comes in. When I read her manuscript for "Mobile Apps: The Ultimate Guide for Small Businesses", I thought, "At last! Someone who speaks to the small business owner about this hot technology."

Cheryl has broken down the barriers to building and launching a mobile app for the small business. And she goes beyond just offering tips but genuinely wants to help this niche succeed. Cheryl served our country in the military because she wanted to make a difference. We're lucky that she is extending that same desire to serve small businesses everywhere who feel challenged by today's competitive marketplace.

Mobile technology is here and it's staying. You can ignore, fear, or reject this reality. Or you can embrace it. With this book, you'll learn how to harness the powerful return on investment with a mobile app. And you'll capture all the extras that Cheryl is offering, including the Mobile App Resource Center and a year's worth of PUSH notifications, already written for you.

I hope that once you finish reading this book, you'll take action. Make it a priority to create a mobile app. There's no longer any reason to wait.

—Raymond Aaron,
Author of *Double Your Income Doing What You Love*

INTRODUCTION

"Nearly half a billion smartphones were sold worldwide in 2011 alone. 67 million iPads were purchased in its first two years in the market; that's the largest product launch in history." According to Flurry Analytics, a platform shift took place in 2011—for the first time, sales of smartphones and tablets eclipsed desktop and notebook sales. For the first time since the launch of cell phones, people are spending more time on mobile devices—and using mobile apps—than the desktop platform browsers.

People are on the go and they need technology that keeps up with them. But the device is really just one part of the connection. These lifelines must also have the software applications that connect them to the functions they need and want—from finding the closest gas station to battling Angry Birds, from checking their stocks to checking their dinner reservations. They're reading product reviews on their phones while standing in a store, just to make sure they're buying the right item, and for the right price.

If you have a business, then you need to pay particular attention to this shift. It's not a trend, but a move toward total mobility, and that means you have to keep up with your customers like never before. If you're not moving ahead with mobile apps, you're already falling behind.

Way behind.

Meanwhile, your competitors are jumping on the mobile app wagon, maximizing their results, grabbing market share, and building customer relationships that create loyalty and retention.

Wouldn't you rather be the business that drives enormous sales with these white-hot mobile apps?

And if you think you're covered because you have a website, think again. Your customers are not sitting at a computer browsing the Web to find you. They're mobile. And your site isn't designed to be viewed on the small screen. So they find one that is more smartphone-friendly, and you've just lost an opportunity to gain a customer and close a sale.

Nor are they opening and reading all the unwanted mail in their emailboxes. Instead, they're connecting with those businesses that understand and respect that their time is precious, and don't inundate them with generic offers and unsolicited information that is clogging up their mobile devices. Instead, the mobile savvy businesses use push notifications to a customer's phone to keep them updated, which they've opted into via that company's mobile app.

I decided to write this book because I have encountered many businesspeople who lack the knowledge and confidence to join the mobile app revolution. In these pages, you'll find a wealth of information and step-by-step guidance to move you toward the essential launch of your business's mobile app. I've also extended the value you will receive here by providing four additional bonuses on my website www.thebookonmobileapps.com.

I've invested my career in studying the best—and worst—practices of mobile app technology, voraciously devouring knowledge and doggedly pursuing the movement and evolution of this rapid-fire mobile world. My goal is to keep my clients up-to-the-minute on the best methods for harnessing the new technology, and I'm sharing my well-stocked inventory of knowledge here with you in this book.

You are going to learn the most cost-effective ways to have a mobile app developed for your business, an app that hones in on the needs and desires of your customers, to better serve them and grab the sales you've earned. You will discover how to avoid being caught up in the shiny object that is new technology and instead focus on the results it can deliver, when executed correctly.

Some people are fine with the status quo, merrily going through each day making enough sales to sustain their bottom lines. But that's not good enough for you. I know this because you've picked up this book. You're not ordinary. You're enlightened. And you want to know how to integrate a mobile app into your business the right way, because if your mobile app doesn't generate revenue, it's not worth the investment.

Let me start you out with three basic rules for your mobile app:

1. Establish a crystal clear objective of what you want to achieve through your mobile app. "Increase sales" isn't defined well enough. "Increase sales from existing customers by 15%" is more like it. Are you looking for brand awareness and social media engagement? How much? From whom? Once defined, don't lose site of your core objectives.
2. Support the success of your mobile app by including the promotion of it in your marketing plan. That mobile app isn't going to download itself. Market to all those people who go shopping on app stores, as well as to your current customers. An app store is a great lead generator, but you have to be there first.
3. Integrate the power of push notification in your app. It's a simple step, but too often overlooked. And given that push notifications get a 97 percent open rate, you can't deny their value. Reach out and pull in your customers when you want, where you want. Nurture the relationship and build your fan base.

You can do it all with a mobile app. And I'm here to help you cash in on this mobile explosion. Before you proceed, be sure to grab those bonuses at www.TheBookOnMobileApps.com. You are going to receive the most incredible free gifts ever.

Bonus #1: The PUSH Motivational Calendar. Do you ever find yourself at a loss for content to share? Not anymore, I compiled one year's worth of motivational quotes and inspirational sayings formatted in 128 characters or less ready to be sent through your mobile app.

Bonus #2: 10 Free Marketing sites and resources to free up your time and triple your productivity!

Bonus #3: A Mobile App Worksheet to guide you through the steps to prepare your business for a mobile app.

Bonus #4: A Mega Bonus, an incredible opportunity to become a Mobile App Insider—For FREE! Let's get going and make a big bang!

CHAPTER ONE

Past, Present, and Future of Mobile

"How lovely to think that no one need wait a moment. We can start now, start slow changing the world."

—*Anne Frank*

The speed of technology is dizzying. It used to be that you could feel confident when buying a computer that it would fulfill your needs for many years, which justified the big investment. Now, the price has come down from those earlier years, but the lifespan has been reduced as well. The average life expectancy of a computer is two to five years.

That is, if you even want to buy one.

From the gargantuan mainframe computers of the 70s to the personal computer of the 80s to the laptops of the 90s and the 21st century evolution to netbook, tablet and smartphone, computers keep getting smaller, faster, and more powerful. And cords have become history as wireless rules.

A Brief History Lesson

I don't want to date myself, but I remember the introduction of the first corded remote for a VCR. For you Millennials out there, that was the machine that played those bulky tapes—not to be confused with an eight-track player, but that's another distant memory.

Long before that, however, there were mainframe computers that were larger than some of today's cars. They occupied their own rooms in the corporate landscape, and were maintained in a practically hermetically sealed, clean environment—a far cry from the laptop on the coffee table that my kid was using as a placemat.

Even during this Dinosaur Age of computing, dating back to the 1950s, the Internet was evolving. But technology wasn't progressing as quickly then. If we couldn't put a man on the moon, could we connect them here on Earth?

Not yet, but soon.

From the behemoth mainframe came the smaller computers. In 1982, technicians figured out how to connect the mainframes to these desktop computers, and the concept of a worldwide network was born, known back then as Internet Protocol Suite. Although it continued to be fine-tuned over the next decade, the Internet didn't really take hold in the commercial/consumer sense until the mid-90s. Since that time, the Internet has grown from handling about one percent of two-way communication (e.g., phone calls, emails) to about 97 percent of the total volume—with the addition of text messaging and video calling to the mix.

The Phone With the High IQ

The technology for a mobile phone isn't exactly new. Back in 1910, Lars Ericsson installed a phone in his car. When he wanted to make a call, he drove to town where he could connect the wires to a telephone pole.

In 1947, Bell Lab engineers Douglas Ring and W. Rae Young proposed the technology for mobile phones while another Bell engineer, Philip Porter, came up with the idea of the first cell towers. It wasn't until 1960 that Bell introduced the first fully automated mobile phone systems for vehicles—which had extreme limitations for service. You found a place near a base station and stayed there for the duration of the call. It was mobile, but barely.

In 1973, Motorola researcher Martin Cooper placed the first phone call using one of those large analog mobile phone prototypes that seemed sleek then but would be considered clunky today. Thus began the long race between Motorola and Bell Labs to produce the first mobile phone.

While they raced to develop the hardware, the first cellular network was built in Chicago in 1977 and activated in 1978. Within the first year, it accumulated more than 1,300 customers.

Once the early models pulled the plug on phones, technology raged on. The second generation (2G) moved away from those large mobile phones and also introduced prepaid mobile phones—largely in response to the complaints about the high cost of mobile calling. The proliferation of cell sites also provided more coverage, making it easier to use the phone in more locations.

In 1993, IBM came out with the first smartphone, the Simon, which included a calendar, address book, clock, calculator, note pad, email, games, and faxing capabilities—all accessed via a touchscreen. Nokia and Hewlett-Packard followed, and the concept of a multi-function mobile device sparked a voracious desire among consumers who clamored for the latest technology. The Palm OS Treo and BlackBerry kicked up the functionality in the early 2000s, but it wasn't until Steve Jobs and Apple launched the iPhone in June 2007 that the phone went from smart to brilliant. Whipped into a frenzy by Apple's pre-launch advertising campaign, customers lined up at Apple Stores—the only place to buy an iPhone at the time—in the wee hours before the retailers opened and paid up to $800 for the privilege of being one of the first to own the amazing device.

The iPhone set the standard for smartphones, but many other companies created their version. The Google Android and Samsung Infuse are popular competitive models.

To keep up with the overwhelming demand for more connectivity, cell technology continued to grow. Fast wasn't fast enough—and probably never will be. 4G was born in 2009 and this network is expected to reach 150 million subscribers worldwide by 2013.

Where the @ Came From

Electronic mail enabled people to send messages back and forth via their computers and actually predated the inception of the Internet. Back in 1965, the geniuses at Massachusetts Institute of Technology sent messages through the school's mainframe system.

In 1971, a computer engineer named Ray Tomlinson, who was hired by the U.S. Department of Defense to develop the first internet—the Advanced Research Projects Agency Network (ARPANET)—sent out the first email message:

QWERTYUIOP

Great moments in email history

- 1971: Ray Tomlinson sends first email message.
- 1976: Queen Elizabeth II is the first head of state to send an email.
- 1982: Scott Fahlman invents the "smiley" emoticon.
- 1989: AOL members hear the phrase, "Welcome! You've got mail!"

OK. Maybe not as profound as "One small step for a man" but he probably wasn't thinking that he was making history.

Tomlinson is credited with coming up with the idea of using the "@" symbol to indicate the location of an email user.

He probably never imagined that email would become the preferred method of communication for many years—and that this little symbol would no longer be just a tiny symbol on a keyboard. And could he ever have guessed that future emailers would step away from their desktop and use this tiny mobile device to send and receive their messages?

The Business Side of the Web

In the not-too-distant past, you were considered hip if your business had a website. Then it became mandatory because customers expected you to be visible online. If you didn't have a website, you were losing leads, customers, and sales.

And these websites created a new way for people to shop. Electronic commerce—e-commerce—is the buying and selling of products or services over electronic systems, like the Internet. The 'Net provides the connection while the World Wide Web connects any online visitor to find virtually anything they want or need, without ever leaving home. Virtual marketplaces, like eBay and Amazon, spawned a new type of retail, but also motivated the brick-and-mortar stores to add an online component.

When online access went mobile, so did the shopping. Mobile commerce—m-commerce—provides the user with the ability to conduct transactions using a mobile device, like a smartphone or tablet. And it has evolved from simple SMS (text messaging) systems to m-commerce applications that make it easy to make a purchase wherever you are, using that device. Add in such up-to-the-second features like located-based services, barcode scanning,

QR codes, and push notifications, and these apps are capturing the attention of consumers who want to get the best deals and the latest product updates to make their shopping experience easy and fruitful.

App Store Aptitude

After giving the world the iPhone, Apple recognized that its brilliant smartphone could get better. Initially, the company guarded its applications development with the same closed doors as its operating system—something that has drawn equal amounts of praise from Apple fans and criticism from the universe of PC users. But shortly after, the company opened its doors a crack and announced that it would allow third-party developers to create iPhone apps, albeit under the scrutiny of Apple's strict rules. In July 2008, just over a year after the first iPhone hit the market, the Apple App Store opened its virtual doors, stocked with 500 apps. In the first weekend alone, iPhone users downloaded 10 million apps. Within a year, the number climbed to 100,000 apps in its inventory, the first app store to hit this mark, and as of February 2012, the store contained more than half a million apps.

Not long after the Apple App Store launch, Google opened up Android Market with 50 apps, but was up to 2,300 apps in March 2009. By February 2012, the number soared to 450,000 Android mobile applications.

So the common website is no longer enough to do the job. If your site isn't designed to grab shoppers on the go, you're leaving money on the table—which your competitors are likely scooping up with glee. If you're not creating a way for customers to interact with you in a social forum—like Facebook and Pinterest—you're missing the conversation that drives visitors to refer, repost, and retweet your name to other buyers.

Your customers aren't just sitting in front of a computer. They are going online with their mobile devices. They take their cell phones everywhere. The checklist before leaving the house is wallet, keys, and phone—and not necessarily in that order. At some point, you will be able to leave your wallet at home, because you'll be paying with mobile apps. And by 2015, we expect that smartphones will have the ability to not only let us know when we're approaching a favorite place, but also offer to pay the bill (with your funds, of course!).

Handbags and briefcases are designed with a place to stash your smartphone, because the manufacturers recognize that this feature is key to the consumer. Cars are now designed with not just an outlet to plug in your phone but a place to store it for easy, hands-free access while driving. Any accessory associated with being on the go is fitted with a place to stash your cell phone.

This is how important phones have become. And the necessity will increase. Those who stay on top of the rising wave will ride high. If not, you'll sink. Don't sink, stay informed, check out even more content and grab your bonuses at www.TheBookOnMobileApps.com.

Where do you want to be?

CHAPTER TWO

7 Reasons Your Business Is Craving a Mobile App

"Whatever keeps you from reaching your goals today had better be important—it's costing you a day of your life."
—Dr. Nido Qubein

You might not be convinced (yet) that you need to develop a mobile app for your business, but if your business could talk, it would disagree. Like the Venus flytrap in "Little Shop of Horrors", it would be begging, "Feed me, Seymour. FEED ME!"

So, before you put this idea on the back burner—and let's face it, nothing worthwhile happens there—here are seven reasons that you need a mobile app for your business.

#1. You Need to Generate, Capture, and Convert Leads.

You can't survive on the business you currently have. If you're not generating and cultivating leads, then you're sitting in stagnant water. You can choose to use the old-fashioned way of digging around to uncover leads, sending out mailers, shaking hands at trade shows, and cold calling—or you can join the 21st century and let technology stimulate a more cost-effective, productive lead generation process. The investment in a mobile app will deliver a powerful return on your investment, particularly when you weigh in the cost-per-lead of your current system.

Do one thing right now. Go to the app store (e.g., Android Marketplace, Apple App Store) on your smartphone or other Internet-connected device. Type in keywords that someone looking for a business like yours would use. See what comes up. Are your competitors there? If so, they've got a head start on you. If they're not there, you have the opportunity to break new ground and reach out to your marketplace before them.

The app stores are an untapped lead generation source that businesses are starting to discover. These m-commerce sites are search engines for smartphone functionality! When you have your own mobile app, you upload it to the app stores and input your keywords to help users find your app (just like you would do to drive Web surfers to your site). You also enter a keyword-rich description of your app in this store, another way to bring those buyers to you.

These app stores are such great search engines that I have seen my clients generate more than 500 new subscribers or hot leads within the first week of going live with their mobile apps. How do I know the leads are hot? The people who downloaded them actively searched the app store to find what they wanted and needed. They sought out the app, so they are clearly interested in what it offers.

Now that you have the leads from people who have your app, cultivate them.

- Tempt them with content that is relevant, timely, and appealing. Dangle a free report, informative newsletter, or demonstration video in front of them—something that hits their pain point and makes them really want what you have. All they have to do to receive this freebie is opt in with their email address. And they will if your enticement is enough to make the swap. And remember that unlike emailings, you're not giving them spam. The app user has the option of requesting this information, and when they do, you have an even hotter lead—and one that is open to receiving this kind of information.

- Use your app to link the user to your Facebook or other social media page, where they can find more reasons to stay connected with your business.
- Send push notifications to your clients and customers. Hitting them on the smartphones has proven to deliver 97 percent open rate. Where else do you get results like that?

And think about this. The shoppers in the app stores have their credit cards on file there, to make purchasing one-click easy. As you seek to monetize your app, you have a captive audience that is not only ready to buy, but can do so in an instant. Get them motivated and when you convert, the customer just has to push a button to feed that desire.

To push your customers to your app, include a QR code or URL in your marketing and promotions so that potential app users can easily find you. It's all about visibility, and when you go mobile, you create more ways than ever for customers to see you.

#2. You Must Differentiate or Drown in a Sea of Sameness.

If you keep doing things the way everyone does, you won't create a buzz. You won't be noticed. You'll prove to the world that innovation is not in your DNA. Stop blending in with the crowd. Exit that field of sameness and step into the bright, shiny, golden opportunity that is mobile apps.

For example, about ten years ago, supermarkets starting converting some of their checkout lines to self-serve. Consumers could bypass lines, scan the barcodes on their items, and bag them (with the bread and eggs on top, thank you very much). It sounded like a great idea!

But the usage peaked at about 22 percent. By 2010, only 16 percent of supermarket purchases were made via self-serve checkout. Sure, some of the customers were intimidated by the technology, but the truth is, the majority of supermarket shoppers want service. They don't *want* to handle the task of checkout themselves.

Customers want service. They want you to make it easy for them to get what they want. It's *your* job to figure out how to make that happen. You need to create a "Wow!" factor, which requires innovation and strategic thinking, a sort of melding of the right and left brains. Look at the present and then envision the future. Where is commerce going? It's online. It's mobile. With time as a premium, as well as fuel prices, shoppers need a reason to go to a brick and mortar location. If you're in this group, you can't rely on traditional advertising because more and more buyers aren't seeing it. They spend 23 hours a day with their smartphones at hand, so why wouldn't you use this technology marvel to your advantage, particularly when mobile apps make it so easy?

#3. Mobile Apps Are White Hot Right Now and You Need to Capitalize on the Momentum.

Just like people eventually had to succumb to buying CDs to replace their records and cassettes, and DVDs (and BluRay) to replace their VHS tapes, the time will come when you will have to make the move to a mobile app. Do it now before you're standing in line behind the other latecomers.

Research by Flurry Analytics of mobile device users showed that people are spending more time navigating on mobile apps than searching the Internet on their desktop or laptop computers. As of June 2011, they were spending an average of 81 minutes a day. By January 2012, that soared to 294 minutes a day. That's a 263 percent increase in just six months. Where will it be in another year?

Flurry also reported that approximately 1.2 billion mobile apps were downloaded in the last week of 2011—yes, just one week! Wouldn't you love to be among the apps that consumers are grabbing?

The numbers will keep rising as more and more people upgrade their mobile devices. Apple's iPhone and iPad reign supreme in this arena but Silicone Alley Insider says that Google's Android activation rate has reached 700,000 *per day*—the equivalent of 250 million activations per year. The opportunity is there. Are you going to let it slip past you and fall in the lap of someone more willing to seize it?

#4. More Smartphones and Tablets Are Being Shipped Than Desktop and Notebook Computers.

I mentioned the conversion from VHS to DVD and BluRay. Well, companies that are still focused solely on traditional websites are committed to the VHS version of online marketing. People *use* computers but they are *enamored* of their mobile devices! Who wouldn't be? These things are always there for you, helping you make dinner reservations, find the best price on gas, getting your theater tickets, organizing your day, delivering your newspaper, videotaping your kid's Little League game—all from the palm of your hand, any time, anywhere.

And then there's the computer on your desk. You can't take it with you. It's connected to the wall. Not too exciting.

Laptops are great, but not as compact as the smartphone. You can't tuck a laptop into your pocket or purse. Notebooks are heading in the right direction, but have been dwarfed, sales-wise, by the smaller devices.

Studies show that 89 percent of your clients are using their smartphones throughout the day—all day, every day. Look around you. How many people have a smartphone in hand? How many are tapping away on the touchscreen versus talking on the phone? Imagine having your logo icon on that touchscreen, next to the likes of Amazon, Facebook, Twitter, and Angry Birds. *That's* branding!

I was watching an episode of "Anderson" recently, with Danny DeVito as a guest. During this segment, DeVito—known for taking a photo of his "troll foot" wherever he goes—photographed his bare foot and tweeted it to his followers, right there from the set. For the remainder of this "interview", DeVito was tweeting. He was answering Anderson Cooper's questions, but his eyes were on his smartphone.

The "phone" feature of the smartphone has become less important than the ability to access other features, like mobile apps. If you want to reach them, connect here.

#5. There Are 1.3 Billion Email Users But 5.2 billion Mobile Phone Subscribers.

Can you smell the opportunity here? Mobile phone subscribers outnumber email users four to one. There are three times as many smartphones being activated around the world every minute than there are babies being born.

This is not a trend. It's a cultural change. Mobile phone usage represents that largest shift in consumer behavior over the last 40 years, easily outpacing the PC revolution of the 80s and the Internet boom of the 90s.

Mobile technology isn't the wave of the future. It's here. Since 2007, more than 500 million Apple and Android smartphones and tablets have been activated. By the end of 2012, Flurry Analytics predicts that this number will surge past the one billion mark. Sitting back and ignoring this opportunity to access the massive numbers of mobile technology users is costing you money every single day.

#6. Your Customers Multitask, Creating More Opportunity to Use Mobile Apps.

Don't take my word for it. A 2011 study by Google, "The Mobile Movement: Understanding Smartphone Users", showed that 72 percent of us use smartphones while also consuming some other media, like watching television or listening to music. Just look around you. How many of you have your phone in hand or nearby when reading a magazine? When you're in a movie theater, do you keep your phone on? Are you texting or playing Words With Friends while all that pre-previews stuff is on the screen? Yeah—me, too.

With the advent of tablets like the iPad and Kindle Fire, digital copies of books and magazines are growing in popularity, creating yet another mobile digital avenue.

This rise in multitasking to include a computer or video screen along with a mobile device is called the "second screen". While viewing one screen (e.g., the television), a viewer is also looking at the smartphone or tablet (the second screen).

Marketers are increasingly exploring ways to complement the first screen viewing experience to drive it to the second screen. They're using QR codes and voting survey polls to prompt viewers to switch over to the second screen (where Internet connectivity can bring them to you).

Undoubtedly, there's a fascination with mobile technology. Users excitedly share the latest app they've discovered. Strangers share hints and tips with one another on the subway or in checkout lines. The question is, will they be talking about you? Maybe. But if you don't have a mobile app, the answer is a resounding, "NO!"

#7. Over 100 Billion Push Notifications Have Been Sent—With a 97 Percent Open Rate.

Mobile apps can have this wonderful feature, called push notification. That means you can send a message to a user even when the app isn't opened. Alert them to a special offer. Give them a quick link to download a free white paper. Push notifications have a 97 percent open rate, versus four to nine percent with email. With your mobile app, you can focus on higher quality leads, avoid the obstacle of spam filters, and bypass the double opt-in process because when a subscriber downloads your app, they get the choice of whether or not to accept push notifications. Like everything else mobile, it's fast and easy for the user—which makes them far more likely to connect with you!

The possibilities are endless. And as advances continue, you'll have more and more ways to leverage the mobile connection with your customers. But you have to start somewhere and sometime. What are you waiting for? Feed me, Seymour! Find out if your business is ready with my Mobile App Worksheet at www. TheBookOnMobileApps.com

CHAPTER THREE

So You Think *Your* Business Is Different?!

"Change your strategy, change your results."
—*Jim Mathis*

There are some times when you need to think like a business owner, and others when you need to step into a marketer's mindset. In my humble opinion, your success requires more time spent in the latter.

If you own a sandwich shop, that's a title. Your competitors have the same title after their names. That doesn't make you different.

When you're the marketer of a sandwich shop, that's a role. Marketers think differently. They envision change and seek out a way to make it happen. They recognize that a business owner creates the availability of a great product or service, but it's the marketing mind that generates desire, visibility, and those all-important leads. Without intelligent and creative marketing, you're just another business.

Like Michael Gerber, author of the bestselling *"E-Myth"* books, so wisely advised, it's time to stop working *in* your business and start working *on* your business. Step back and see your company for what it is and what it can be. When you open your mind to possibilities, you'll be amazed at the greatness that saunters in.

Think Strategically, Not Tactically

A marketer thinks (or should!) strategically, uncovering opportunities and creating ways to maximize them. You need to adopt this manner of thinking and step away from the taskmaster mentality. With strategic thinking, you focus on solutions, while the tactical brain is concerned with the tasks that need to be handled throughout the day. Those chores will only slow you down and distract you from your more important focus. You'll feel overwhelmed as your "To Do" list mounts. That leads to procrastination, which begets frustration, as you're seeing no results. Frustration will take you to anger and, at that point, you might as well close down shop because nothing useful will happen.

This is the place where the majority of "business owners" are stuck. But there's a way out if you can adjust your thought process.

There are two types of strategic thinking:

- **Strategic synchronous**: Constantly focused on what the competition is doing for marketing so you stay aligned (i.e., in synch) with them.
- **Strategic continuous**: Ignoring the competitors' marketing and charting your own course with an open mind to opportunities and possibilities.

As the marketer (not owner), you should spend your time looking ahead, identifying avenues (niches, media, promotions) that your competition hasn't explored. Leave the "doing" to those people in your organization who are better suited to taking you from Point A to Point B. Your role is to pinpoint where Point B is going to be located.

Cheryl Paulsen

The Answers Are Not in Your Industry

The best place to find new ideas for marketing your business is not uncovered by following your own industry. Anything you do here will fall into the "been-there-done-that" category of uninspired thinking.

If, for example, your competitors traditionally advertise in the Yellow Pages, you have to ask yourself, "Does this work?" I like to follow the timeless mother challenge: "If Johnny jumped off a bridge, would you?"

Just because the other businesses in your industry take a certain tactic doesn't make it right for you. Be an individual. Look **_outside_** your industry. If you're in the automotive business, look at successful restaurants and analyze how they attract customers and keep them coming back. What are other businesses doing that could possibly work for you? Could the QR codes that Starbucks puts on a coffee cup be a good medium for a mixed martial arts studio to generate leads? How can you integrate new or different technology—like mobile marketing—to your mix in a way that will deliver results? If you were to launch a mobile app, what should it be able to do for your customers?

Look at direct marketing. Some people say it's dead, but the statistics show that the report of its demise is entirely premature. I run an Internet-based business, but I can appreciate the fact that people who purchase my services have mailing addresses where they receive actual mail. Why should I assume that the only way I should reach out to them is via the Internet? That would be grossly short-sighted.

Harley Davidson broadened its scope from conventional marketing venues for brick-and-mortar stores. Instead, they observed how Tupperware Parties brought customers together in a social environment that prompted purchasing. The motorcycle company then instituted "Harley parties", which capitalized on the strong brand loyalty and brought together loyalists. Their gamble paid off and Harley Davidson's marketing changed the dynamic of the motorcycle buying process.

The best way to build a relationship with the customers you want is to connect with them via multiple channels, both on—and offline. Different media has its own unique benefits, so consider each one and how it will hit your target.

Be Your Clients, Not Yourself

Your clients are human beings, just like you, but that could be where the similarity ends. Don't let yourself make the fatal mistake—believing that your needs, desires, and perceptions mirror those of your customers. Don't force them to fit your model of the ideal customer but be aware of who they are. Why would they want your product or service? What do they value? What influences their buying behavior and decision making?

I've worked with many clients who don't own smartphones or tablets, but they recognize that their customers do. It's often hard for them to understand things like push notifications and multi-text messaging, but they don't need to embrace this technology. They just have to recognize that it's an excellent channel for communicating their marketing messages to their customers who are enamored of their mobile devices.

Reach out to your customers in the channels that they use. Do they get their news from television, online, or a mobile app, like Newsstand? Are they likely to respond to pop-ups or pay-per-click ads? If they aren't an email bunch—like Generation Y—then mobile technology and social media marketing make sense.

When you're ready to differentiate yourself from the competition, start by talking to your customers in a better way. Show them that you are in tuned with their needs and interests, beginning with the way you communicate to and with them.

Be More Than a Provider

We are all striving to become hugely successful. I want to make more money, work less, and create more time to spend with my family. I want to take control over my time, not vice versa.

And I have to believe that you aim for similar goals, or you wouldn't be reading this book. You're seeking ways to generate revenue and you know that mobile marketing is a great place to look.

You also know that, no matter what the economy, there is some business, some industry that is thriving. Look at the automotive industry. While American automakers were struggling to stay afloat, Ford Motor Company found a way to outshine them. They developed the types of fuel-efficient but fully appointed vehicles that American consumers want. They learned how to communicate with their clients in effective ways, building a community of Ford lovers, not just a string of customers. Ford recognized that people don't just buy cars, they want an experience—and they want a brand that echoes those desires and reflects their values. And Ford, by the way, has launched several mobile apps to give their car owners more information, entertainment, and—above all—a stronger connection with the Ford brand.

Whether you choose to develop a mobile app for your business or not, recognize that your customers are looking for guidance. They might not communicate that need, but it's evident in the choices they make. As Steve Jobs said, "People don't know what they want until you show them." They choose products and services that deliver a solution. They want to save time with more convenient options. They want choices, quality, and value. They don't necessarily know what that is or where to get it, so it's your job to be highly visible with the message, "Here it is!"

Your goal with your marketing should be to solve problems for your customers. Be their trusted advisor and reliable resource. When you achieve this, you keep them coming back to you and strengthen the bond. It becomes easier and easier to sell to them because they respect your brand and what it stands for.

Nix the Naysayers

Now, here's the caveat. Once you rise above the competition, they'll be taking pot-shots to bring you down amongst them. That's a true sign of success.

Look at the sports world. Back in 2001 when Tom Brady led the New England Patriots to the team's first Super Bowl victory ever, in his first season ever, he was the golden boy and the Patriots soon became "America's Team". By the time they won their third Super Bowl in four years, the Pats were no longer underdogs but became the object of resentment. That's what comes with success. You have to recognize envy as a form of flattery, develop a thick skin, and don't allow the critics to cause you to question your methods.

At the same time, recognize that what elevated you to this height was—and is—your innovative vision. That's what separates you from the pack. While they grumble that, "This is not how we do it in this business," you know you're doing the right thing. By not following their example, you differentiate yourself, your company, and your brand.

Cheryl Paulsen

Deliver the "Wow!" Factor

There was a time when innovations were few and far between. Now we're always waiting expectantly for the next big thing— that "Wow!" factor. Apple has been a brand that delivers this pizzazz, from the iPod that revolutionized the way we listen to music, to the iPhone that put the "smart" in phones, and the iPad that turned mobility from something cool to something we can't live without. That was real "Wow!"

Netflix and Redbox both delivered "Wow!" when they provided brilliant solutions to movie rentals, overcoming the mind block that derailed Blockbuster.

What can you do to grab the attention of your market? What "Wow!" will make you a stand-out? When you identify this factor and serve it up to your customers throughout your sales cycle—including the mobility side—you cultivate followers that will continue to propel your success. Get even more ideas with a free membership to the Mobile App Insider's Circle, www. TheBookOnMobileApps.com.

It's fine to say that your business is different. I challenge you to prove it.

CHAPTER FOUR

5 Mistakes Small Business Owners Make Building An App

"Determine the thing that can and shall be done, and then we shall find the way."

—*Abraham Lincoln*

As a small business owner myself, I see that some others don't know how to effectively get their message in front of their targeted market when they want to reach them. Email services are great, but this medium has its limitations—spam filters, cluttered inboxes, outdated addresses—that contribute to a low open rate.

Social media marketing is exciting and everyone wants to jump on board, but this medium isn't a goldmine either. Not all of your customers are on Facebook, Twitter, LinkedIn, or Pinterest. And those who are can easily miss your message if they don't "Like", "Follow", "Connect" or "Pin" you. And they also have to be online at a time when they can see your update. I love social media for many reasons, but it's not a panacea for today's businesses.

I know that one of the biggest obstacles owners face when considering mobile app development is that it seems too big, too complicated, and too much work to manage. They fear that they'll need to add an IT person to the payroll just to manage this new asset. They don't understand what it takes to get an app built and launched. So, they either do nothing or stumble through a series of missteps on the way to getting it done.

In a global marketplace that is further challenged by a tough economy, it's more important than ever to differentiate yourself. You need to take risks and do something that the others haven't tried. "Business as usual" has put us squarely in the biggest economic mess we've experienced in decades. If you don't keep your head on a swivel, constantly scanning for new ways to be different, you'll wind up as just another boarded-up shop and abandoned building. "Business as usual" is complacency, and complacency is failure.

The good news is that the future is looking mighty bright for business owners like you—people who are committed to doing what it takes to succeed. The visionary who will design a recession-proof business that will rise above the current and future economic storm clouds. The customer-centric entrepreneur who will communicate with buyers via their chosen channel: mobile devices.

A mobile app will undoubtedly give you a competitive advantage. This tool will provide instant access to reach your clients when you want, where you want. You've already set the foundation by creating your website. Now it's time to stop waiting for them to find you there but to cast a net and pull them in.

As with any technology, you'll encounter pitfalls. If you can navigate them, you will be leaps and bounds ahead of the competition, saving yourself days—if not weeks—of headaches. So, here are the five mistakes that business owners make when stepping into mobile app development.

Mistake #1: Not Having a Clear Objective or Strategy.

Don't lose sight of your core business objectives. Be clear about what you want to achieve. For example, are you looking for sales, brand awareness, social media engagement, or VIP rewards? Once you establish your desired outcome for your mobile app, it's much easier to stay on track and develop your key strategies, using those goals as your target.

Think back to all of the projects in which you have succeeded and those where you didn't reach the outcome you wanted. It's just as important to look at our mistakes as our wins. A failure is actually an opportunity to learn.

When I begin a project with a client, we proceed with the end fixated in our minds. By doing that, we're able to make good decisions throughout the processes of building, marketing, and launching the app. This clear objective keeps us on the path to success.

Now, you might be saying to yourself, "Well, that's just great, but I have no idea of what my objective is. I've never built a mobile app before. I've never even considered getting into this arena, so how am I supposed to come up with expectations and strategy without being taken advantage of by someone who can see I'm unclear?"

That's a fair concern, and one I've heard quite often. This is where working with a professional mobile app designer or developer will help. Just like a webmaster can guide you through the possibilities of a website, a mobile app pro can help you create structure and identify features and opportunities to pursue. You probably have yet to explore the massive potential of a mobile app so this is a good time to check out the options. This important step will provide you with valuable insight to formulate your objective.

This is also a great time to re-evaluate your unique selling proposition (USP). What truly sets you apart and strengthens your appeal to your customers? How will that USP resonate with mobile customers? When these consumers are on the go, what can your business offer that will draw them to connect on their mobile devices? Now put that message at the very top of your objectives.

Mistake #2: Thinking You Don't Need a Programmer.

Most small businesses don't require a completely customized app, created from the ground up. It's easy to get carried away with all the cool features that are available. It's also easy to see a cost estimate upwards of $10,000. Don't gasp! Knowing and focusing on your objectives will help you lower the price.

Hiring a programmer is helpful because you're a marketer of your business, not an app developer. When you need your teeth fixed, you go to a dentist. When your plumbing bursts, you find a plumber. When you want to build a website, you hire a web designer. It just makes sense to hire a professional to avoid costly mistakes.

There are many different ways to find the right programmer for your mobile app. You can hire a full-service, design-and-build agency or seek out professional freelancers on sites like Elance.com and Guru.com. These sites allow you to post a detailed project description and solicit bids from thousands of registered service providers around the world. Whatever you choose, be sure that your specifications are clear in your description and that you are equally clear when reviewing the bids you receive. Do you want cradle-to-grave service that begins with helping you identify the app's objectives? If you don't know what functions you want, ask the bidders to identify what they're proposing and including in their estimates. A bit of advice though: Don't hire someone to build your app on an hourly rate. You can get nickel-and-dimed up to way more than you expected—or needed—to spend.

Mistake #3: Thinking You Can Do It Yourself.

There are plenty of do-it-yourself (DIY) sites out there that will help you develop a mobile app. What they don't tell you up front is that you have to build your own graphics to certain specs for mobile devices. It's not impossible, but certainly not as easy as a plug-and-play version. You should also know that Apple rejects about 60 percent of submitted apps.

The DIY sites offer a fantastic construct for software. The problem is actually getting all the information plugged in there. They might not be as plug-and-play as they seemed on their site. If you don't know your objectives (see Mistake #1) or what the features can do (see Mistake #2), then you are destined to make Mistake #3. The features you build out need to address your objectives, and you need to know what the features can do in order to be sure you're not taking a detour from your desired outcome.

A notepad on the surface might just seem like a notepad, but when you start to put that into different industries, such as mixed martial arts, you can now make your app a Brazilian Jjiu Jitsu journal or a mat chat notes. Now all of a sudden, you've gone from a simple notepad for one to a way for your customers to interact with you, giving them a reason to go back to your mobile app over and over again.

The graphics are also a big challenge with these free software sites. They fail to tell you when you sign up that Apple sets specific software specifications for the iPad and iPhone models. Apple also instituted graphics resolution requirements that will change with the updates to iOS (Apple's operating system). For example, Apple now requires a certain pixel size that will work with the retina scan within the iPad and iPhone iOS. So, before you decide on the app software you will use, be sure that you know the final costs. Be sure to include the monthly hosting fees and the applications that you will be able to use for those fees; iPhone, iPad, Android, HTML5 for BlackBerry.

Also make sure that you're prepared with high-resolution images of your logo or photos that will be inserted into the graphics. You'll probably need some type of photo editing software—like Photoshop or Photoshop Elements—to combine all of your images. You will have to build your home screen, your opaque background screens, your splash screen, and your app icon image to certain specifications in order to upload effectively to the Apple App Store and Android Marketplace for approval.

It's certainly doable, but be prepared for more than a couple hours of work.

Mistake #4: Inaction.

You can take all the time you want to ponder decisions along the way toward launching your own mobile app, but speed of implementation equals success.

Fear is your biggest enemy, and it manifests itself in many ways. Some people fear failure, but others are afraid of feeling inadequate. Fear sparks inaction. Inaction prompts failure. As marketing guru Bill Glazer says, "The only difference between an average income and financial freedom is implementation."

The fundamentals of business haven't changed, but the economy and the environment have changed. The Web has given people access to far more information, so they have become skeptical. This skepticism makes it harder for your customers to take action. You're in a position to take action over inaction. Put yourself in your client's shoes. People are busier. You're busier. It's easy for us entrepreneurs to become overwhelmed with all of our great ideas. The possibilities are endless. But don't let the quest for the "Perfect App" stop you from riding the wave of this amazing tide of opportunity. Failure plus perseverance equal success. Get out of your comfort zone and take action. How would you feel if you dragged your feet some more and your competitor launched a mobile app in the meantime, stealing away your customers who thought it was the greatest thing ever? You know that you'd kick yourself. Save yourself the time. Get moving NOW!

Mistake #5: Not Marketing the App.

Your mobile app isn't going to download itself, just like people won't find your website without you doing some search engine optimization (SEO). You need a marketing plan for the launch of your mobile app. Marketing is the backbone of your business.

You need to think about your target niche and networks as you build your marketing plan. Change the way you think about your business, because that's what your mobile app will do. It will enlighten, excite, and entice your customers in a way that your website cannot. They love the ability to get things done via a mobile device. Give them a reason to love that smartphone or tablet even more by letting them know about your new app. Alert them as to why they need to stay connected through this channel. What will it do for them? Remember the timeless advertising mantra: "What's in it for me?" Tell them what they'll be missing without your mobile app, like previews, specials, and timely updates. Poll them for feedback to show that you're interested in their opinions.

Next, keep in mind that your customers aren't just the consumers or clients, but also your vendors and team members. Include them in your plans to drive a successful launch.

The worst thing you can do is to wait for money to change hands before you begin your mobile app marketing campaign. Be the most interested _in_ your clients and the most interesting _to_ your clients.

CHAPTER FIVE

9 Steps to App Success

"Knowledge is of no value unless you put it into practice."
—*Anton Chekhov*

I've spent four chapters getting you into the mobile app mindset, explaining the importance and return on investment (ROI) of developing a mobile app for your business. You know that if you're not perched on the cutting edge of innovation and technology, you're likely to get your market share sliced by those competitors who are more visionary.

Now it's time to harness those mobile devices so you can unleash the power of the latest technology to build relationships with prospects and customers. Whether you choose to build your own app with the available online software or hire a designer or programmer, follow these nine steps to achieve greater success.

Step #1: Determine Your Objectives and Strategy.

You wouldn't use a power drill to dig a hole. The tool doesn't match the objective. So, before jumping in to mobile app development, stop and think about two critical components:

1. <u>Objective</u>: What do you want your app to do?
2. <u>Strategy</u>: What's the best path to take to get there?

We've talked earlier about your objectives. Now I want you to itemize them. What is your corporate goal? Are you looking to increase sales, expand your brand awareness, deepen social engagement, or build customer retention? Be clear about your destination so you can steer your mobile app in that direction.

As entrepreneurs and small business owners, it's easy for us to become overwhelmed with all the responsibilities that hit us square in the face, day after day. Compound that with the steady onslaught of good—and even great—ideas that demand your attention. It's staggering! At the same time, we can get so deeply mired in the mass of "stuff" that we fail to move in any productive direction. Maybe we are overdosing on multi-tasking or stalled from the desire for consummate perfection. The possibilities for your mobile app are endless right now. Don't let the quest for the Holy Grail of All Apps to prevent you from riding the wave of this opportunity, or you'll miss the crest and drown in the receding tide.

Once you identify your corporate objective for a mobile app, you can feel confident in your ability to decide how to build the app (or have it built). Most small businesses don't need a completely custom-built app, but can modify a program to meet their needs. It's easy to get carried away with all the cool features that are dangled before you, but keep your focus on your objectives. Measure every option against its contribution to that desired outcome. Otherwise, you'll not only get more than you need (or can manage) in your app, but you'll also have a development bill that will exceed your expectations—and not in a good way.

Take your objective and use the information I've provided so far to choose the smartest route to app development. Do you have the time to commit to using a do-it-yourself site? For some do-it-yourselfers, the desire to learn the inner workings of the program is as valuable as getting the app. If you have a good aptitude for working with technology, by all means, give this a go! But also allow yourself an escape hatch if it gets too difficult. Don't waste your valuable time wrestling with a project that would be better delegated elsewhere. Remember the statistics about Apple's 60 percent rejection rate. That hurdle is looming ahead of you. If you don't feel confident that you can soar over it, invest in an experienced professional to give you a lift.

Step #2: Identify Your Most Useful and Engaging Information.

A mobile app is not the same as a mobile-enabled website. The goal is not to require a user to navigate page after page of content. Your app should be quick and easy, designed with the intention of driving the user from your app to your site, not to replicate it.

So how do you do this?

Filter down your messaging to the equivalent of an organized billboard. Communicate the most appealing aspects that will get them engaged with you. Tell them what's new, hot, and exciting. Don't rehash what they already know or just try to get them to spend money. What do you want them to know about your business? Is there a service or product that is under-utilized? How can you present that information as more enticing to your users? What benefit can they enjoy by coming back to your app? Will they earn redeemable points or have a chance to win something? If you want them to keep returning, give them a reason. Build your Know, Like & Trust factor through one valuable tidbit after another.

And do not feed from another site. Apple will hit the "Reject" button if you're just taking your content from elsewhere.

Step #3: Integrate Your Social Presence.

All of your social media pages—Facebook, LinkedIn, Twitter—should be linked to your app, and vice versa. Incorporate those vital links in your app so that people can find numerous ways to connect with your business. Your objective is to build relationships with these customers. The multi-channel approach is the best way for them to return repeatedly. You'll increase your brand awareness, build your credibility as a trusted, valuable resource, and ultimately boost sales.

Encourage your mobile app users to tie their social persona with you, so you can leverage their networks.

- Invite them to "pin" you on their Pinterest boards so they can share what they love about your products with their friends on this site.
- Give them an incentive—such as offering a freebie or hosting a contest—to share your link with their friends, followers, and connections.
- Link to your YouTube videos where they can see entertaining product demonstrations and useful "how-to's".

Pull it all together and create a one-stop shopping and networking experience so you can maximize your investment in both your mobile app and your social media efforts.

Step #4: Create a Loyalty Program.

People who download mobile apps must have a reason to take this step. They aren't going to clutter up their smartphones and tablets with useless icons. Your app needs to fulfill a function for them. And that comes down to the three basic benefits: save time, save money, or just plain entertain. Look at the apps you have on your phone. Zoom in on the non-native apps (the ones you downloaded versus the ones that came with the phone). What do they do for you?

One of the most common and highly successful uses of mobile apps—and especially for small businesses—is offering loyalty and VIP programs. Instead of printing membership punch cards, program your app to track a user's loyalty points and rewards. Create as many as you want or need to maintain your user engagement. Use your new ability to reach your loyalty and VIP members by sending push notifications so you can keep them updated on time-sensitive specials and show them how to earn more rewards.

Step #5: Develop Your App Marketing Strategy.

I've told you before, your app isn't going to download itself! Just like people won't find your website without search engine optimization (SEO), your app will be dormant in the app store if you don't market it to your prospective users. You need to prepare a marketing plan for both the launch and ongoing promotion of your mobile app.

Here are some points to integrate into your planning:

- Send an **email blast** to your database with a URL that links directly to the app on the Apple and Android stores. In your message, tell them the app is available and why they need to download it.
- Print a **QR code** on all of your promotional and corporate materials. The code will link directly to the Apple, Android, or HTML5 version of your app. Put the QR code on signage, packaging, ads, brochures, flyers, post cards, decals, business cards, and any place where your logo is visible. Print it in your digital and published books, eBooks, and white papers. If you mail out invoices, print

the QR code on those forms! Don't miss an opportunity for any possible connection.

- Do **keyword research** before submitting your app to Apple and Android so you know the words that are gaining traction on those site searches. Then integrate those keywords in your app, its description, and in any mention of your mobile app (press releases, ads, web copy) that will appear online. Remember, the app stores represent a new, untapped lead generation source for your business, and good use of relevant keywords will help you to unlock that door.
- **Blog** about the app and link those posts to your company's Facebook and Twitter pages. Make sure you integrate your keywords into the blog post, too.
- **Survey** your customers. Ask those who are using it to provide feedback so you can continue to improve it. This step not only garners valuable market research, but also reinforces your interest in serving your customers—and alerts anyone who has overlooked your app to give it a click.

Step #6: Track Your Downloads With In-app Analytics.

Before you can begin to communicate with your prospects, you must gather an audience. Just like building your email list, you should track the app subscribers and downloads. You can do this with the analytics that are built into the app software. These analytics give you insight into how many people you're reaching when you send out your push notifications. The figures are also critical when you start looking at generating revenue from advertising income. When you can provide potential advertisers with impressive data on your users, you strengthen your market value and advertising rates.

Step #7: Update Your Dynamic Content.

Don't wait for your customers to come looking for you. Your job is to entice them. If a retailer never changed his window display or brought in new inventory, the store would lose customers. The same truth applies to your mobile app. Keep it fresh and dynamic. Show that you're actively looking to excite your customers. Share daily, weekly, or monthly specials. Are you going on the road where they might want to connect with you at a special event? Are you hosting a seminar or going to be a guest speaker on a topic that might interest them? Your mobile app is another vehicle to channel news to your subscribers. When you can keep them interested by communicating the types of notifications they like to receive, you build loyalty that leads to more sales and referrals.

Step #8: Send Push Notifications.

This is a big one. Use your mobile app to send push notifications to users. The notifications show up on the smartphone's or tablet's screens, even when your app isn't opened. Unlike an emailing, they don't have to open it to get the message. It's like a billboard. Grab their interest in passing. Studies show they have their phones with them 23 hours a day so you'll get their attention. Your message WILL be delivered. Use those push notifications to give them a quick update of what's happening in your business. They downloaded your app so they've told you they care. Now, treat them like any lead on your email list. Nurture them to sales—which means you're respectful with the number of push notifications you send. With too many, they might delete you, but when you don't send any, they'll forget you. The balance lies in the middle.

Step #9: Monetize.

You're investing in developing and promoting your mobile app. Make sure you're getting a return on that investment. Increase your earning potential beyond the sales from the customers who downloaded the app. Broaden your vision to selling advertising on this piece of marketing real estate. Use the ad space to promote your own products or sell the spots to companies who are promoting complementary products that would interest your subscribers.

What other businesses might want to reach your mobile app subscribers? If you have a restaurant that caters to upscale diners, you could sell ads to a theater where they might want to see a performance after dinner and a car service to get them there. Offer advertising space to a new boutique in town that is looking for more visibility among your crowd.

If you're selling real estate on your app, your ad space has real value to home supply centers, interior designers, renovators, and all the businesses in your area that want to welcome new homeowners.

The more downloads of viable users you gain, the more your new list will be worth for in-app advertisers.

A successful mobile app is a combination of intelligent content and structure and a finely tuned marketing effort. Build a great app that delivers something worthwhile to your intended market segments. Then make sure they know about it through every means you have to promote it. And when you build that loyal following, reward your efforts by monetizing the site through paid advertisers. Follow the nine steps I've outlined here and you'll discover the true value of mobile marketing.

Finally, keep in mind that people buy from those companies they know, like, and trust. Putting your brand on the mobile devices they already know, LOVE, and trust puts you in the right place and at the right time.

Continue your knowledge and success and be sure to grab those bonuses at www.TheBookOnMobileApps.com. Remember, you are going to receive the most incredible free gifts ever. Bonus #1: The PUSH Motivational Calendar. Do you ever find yourself at a loss for content to share? Not anymore, I compiled one year's worth of motivational quotes and inspirational sayings formatted in 128 characters or less ready to be sent through your mobile app. Bonus #2: 10 Free Marketing sites and resources to free up your time and triple your productivity! Bonus #3: A Mobile App Worksheet to guide you through the steps to prepare your business for a mobile app. Bonus #4: A Mega Bonus, an incredible opportunity to become a Mobile App Insider—For FREE!

CHAPTER SIX

Tips for Apple Compliance

"You have to learn the rules of the game. And then you have to play better than anyone else."

—*Albert Einstein*

Sixty percent of the mobile apps that are submitted to Apple are rejected. That means the odds are against you getting your app into the App Store. But there are currently 550,000 apps across 21 categories parked there, so clearly, acceptance is not impossible.

There's an entire manual on the Apple Apps Developer website. This extensive document outlines the rules for complying with Apple's very specific requirements. With all of the apps that I've developed, I know the errors that can cause you to get rejected. I can now give you the top 14 things to do in order to get approval of your app on the first try.

For those of you who are planning on going the do-it-yourself route to mobile app development, this chapter is a critical read. By following these tips, you will vastly increase your potential for success on the App Store—saving you time and money—so you can join that heralded corps of Apple-compliant mobile applications.

Rule #1: Utilize Tabs.

Incorporate tabs that link to native functions on the iPhone and iPad platforms. For example, if you incorporate an "Email Photo" tab in your app, that's going to include the camera function of the Apple device. When you utilize the "Messages" tab for your push notifications, you are using the native texting function. A QR coupon tab uses the camera function and a GPS coupon tab uses the "Map" function that is built into the iPad and iPhone.

Rule #2: Don't Create a Blatant Sales Tool.

Your app cannot solely be a marketing tool for your business. You can't, for instance, build an app that just lists your services or products. You must be providing a useful function with your app. This distinction represents a fine line for small businesses that are building an app with the intention of marketing themselves. If you want Apple's approval—and connection with their massive network of App Store customers—you need to come up with an effective way to encourage your subscribers to come back to your app over and over again. Not only is that Apple's requirement, but it also makes sense for your business to have this ongoing relationship with your customers.

Rule #3: Include Dynamic Content.

Apple looks for dynamic content within an app. That means you must be regularly adding fresh material—news, product updates, special offers—that entices subscribers to return to your app after downloading it. Ask yourself, "Will someone find the need to open my app more than once?" If not, Apple will reject it.

There are many ways to add dynamic content to your app. Think about things your customers want to know. Is there a new product coming? Are you expanding your business? Have you changed the business hours? Did you add a new collection? Have you received an award or special recognition, like Best Wood-fired Pizza in the City? Think about your subscribers as a group of loyal customers crowded around your lobby or showroom floor. What could you shout out to them to spark their interest? Have fun and be creative. Remember that your app will be reviewed by a human, not an algorithm, so these people will be asking themselves this very same question!

Rule #4: Make Sure It's Complete and Connected.

The quickest way to get a "Reject" notice from Apple is to have broken links or blank sections. And, just so you know, including the words "Coming soon" does not fill in a blank page or tab. In the same way you would (or should) proofread a document before sending it out, check the functionality of every link and review the content on every tab on your app to make sure it's complete.

Rule #5: Don't Mention the "Other Guys".

If you want Apple's approval of your mobile app, do not mention any other mobile platforms. Any reference to BlackBerry or Android, even if you're saying Apple is superior, will get your app booted from consideration. You cannot bash Apple or compare them in any way to other platforms.

Rule #6: Create Multiple, Useful Tabs.

I recommend creating an application with six to ten different tab functions. Only a few times have I seen an app with three to five tabs make it through the Apple review process. If you think about it, coming up with six to ten tabs is not difficult. Here are eight basic ones to get you started:

1. Home
2. About Us

3. Contact Us
4. Facebook
5. Twitter
6. LinkedIn
7. Specials/Coupons
8. Email Us

Rule #7: Keep It Clean.

Lots of kids are downloading apps these days. Some of their devices have parental controls installed, but many don't. Apple needs to know that you're keeping an eye out for kids by including non-offensive content. Keep your app G-rated, which Apple considers anything that is appropriate for a child over the age of four.

Rule #8: Be Unique.

Apple has more than 550,000 apps in its App Store. I can assure you they don't need any more fart simulators or flashlight apps. Suffice to say that if your app doesn't do something useful or provide some form of lasting entertainment that is unique from any apps already there, you're at high risk of rejection. Before you develop your mobile app, do a search of the App Store to see if there's already something similar there. If there is, make darn sure you can differentiate your app with unique, valuable features and functions.

Rule #9: Look Professional.

This is not amateur hour. If your mobile app looks like it was cobbled together in a few days, you'll be rejected. If you think you can start out with a practice app to impress your friends, you'll get rejected. The world is full of serious app developers who don't want their quality work in the same place as noticeably novice programs. Get serious, or get out.

Rule #10: Avoid Controversy.

If you want to make a bold statement to the world, don't expect Apple to give you a platform in its App Store. Apple rejects apps for any content or behavior that they believe to cross over the line of propriety. Where is that line drawn? Well, as a Supreme Court justice once said, "I'll know it when I see it." Use good judgment and stay away from controversial subjects, questionable content, and just, plain inappropriate behavior.

Rule #11: Function With Both the iPhone and iPad.

If you want Apple to accept your mobile app, your program must run on both the iPhone and iPad without modification—and not just the latest, greatest version. You need to program the app for the previous generations of the devices and make the resolution compatible with the iPhone 3GS and original iPad as well as all that came after it.

Rule #12: Size Matters.

Any mobile app that is larger than 20 MB will not download over cellular networks, so Apple is not going to approve it. That's a wise requirement. Just like the seatbelt law is designed to protect you against yourself, this less-than-20-MB is for your benefit as well. If you were allowed to put a huge app out there, your subscribers would be very angry over the time and data it takes, and you would be deleted before the download was finished.

Rule #13: Don't Keep Secrets.

I know of quite a few people who thought they were more clever than the folks who review apps for Apple. Ha! Do you think Apple achieved its global success by allowing cracks in its armor? Any type of secret functionality will immediately get you bounced. If your tab says that it does something in particular, like present special offers, and instead it takes you to a website when you open the link or, even worse, has no special offers, then the trick's on you. Apple will make you disappear.

Rule #14: Get Consent.

Apple demands that you obtain user consent before collecting and transmitting user data or sending push notifications. It's simple to program the permission function into your app, so just remember what your mother always told you and just say, "Please."

If You Get Rejected

Apple has a review board so that if your app is rejected, you can appeal to this group. Like I said, apps are assessed by people, and people make mistakes, so there's a possibility the rejection can be overturned. At the very least, you'll learn why your app wasn't approved and can perhaps make the necessary adjustments to the program to gain Apple's blessing. Just be sure to remain calm and professional. Fight the urge to take your case to the media and attempt to trash the company. Trust me. That's not a wise move. You need the lead generation power of the App Store, so either play by Apple's rules or don't play at all.

If you've had problems in the past, let us know, we can help. Connect with us here, www.TheBookOnMobileApps.com.

CHAPTER SEVEN

Marketing Keys to Success

"We were born to succeed, not to fail."
—*Henry David Thoreau*

You can have a brilliant idea, the next great product, or life-changing service, but if you don't market correctly, it's just unheard of—literally. The same goes for your mobile app. You can put everything right into the development and even be one of the 40 percent of people who gets approved by Apple, but if you don't tell the world about it, the only ones who download your app will be those who accidentally stumbled on it.

Yes, I've already told you that your app won't download itself. Now let's look at how to increase your subscribers with smart marketing.

Marketing is the action or business of promoting and selling products or services. That's the textbook definition. In the real world, marketing is any means you take to draw attention to what you're selling, with the hopes that people will want to buy it.

Direct response is a type of marketing that you should better understand. Just like the name, this form is aimed at soliciting a response—an inquiry or purchase. The response is specific, like "Call now" or "Click here to download." Because you're asking for an action, you can measure the response. For example, you run an ad in the paper with a coupon. You can count the number of coupons that are redeemed to gauge the success of your ad.

Conversely, you can run an ad that has a photo of your product and has your contact information, but NO call to action. You forgot to tell the consumer what you want them to do. Sure, you think it's crystal clear, but with the thousands of messages that bombard a person every day, they can only react to a few. If you haven't asked for an action, don't expect one.

Any ad—whether in print, on television, or online—that doesn't feature a call to action is nothing more than a listing. You fail to create any urgency, to spark desire, or to ignite the need to act. This is the biggest mistake people make when advertising in the Yellow Pages—yes, those directories still exist; mine is a door stop. Advertisers simply list their services and a phone number with the assumption that a consumer will know to make the call—it *is* a phone directory, after all. Now, it's true that most people go to the directory when they have nowhere else to turn to find a resource, but it's been proven that an ad with the words "Call us" generates more response than those that ask for nothing from the reader.

What I'm getting at here is that you can't just introduce your mobile app and expect that people will jump to download it. Give them a reason to do it. Ask for an action. Tell them what they will receive when they download the app.

If It Doesn't Work, Fix It.

How many of you have a website without much traffic? How many don't even know how much traffic you have? If you're not keenly aware of who's looking at your site, how often, and what they are looking at most, how do you separate what's working from what isn't? And if you don't have enough traffic on the site, would you pull the plug on it?

People are more willing these days to put money into their websites. They'll invest in fixing it because they believe this is a good marketing medium. But these same people run an ad just once or send out only one mailer, and unless that effort draws a flood of response, they pronounce that advertising and direct mail just don't work.

The truth is, it's not the media choice that is necessarily failing, but the execution. You can mail a card to a targeted list of buyers, but if the message on the mailer isn't compelling enough, you won't get a response. That doesn't mean that direct mail is dead. Quite the contrary, in fact! Direct mail is experiencing a resurgence right now. With spam filters and overloaded email boxes, the open rate on email is barely four to nine percent. This reality opens the door for other avenues.

You can create a brilliant ad, but run it in the wrong place or at the wrong time. The advertising medium didn't fail you. Your execution failed.

Clear your head of the misconceptions that direct response marketing doesn't work or you will miss a great opportunity.

Four Types of Direct Response

There are four different types of direct response:

1. **Direct Sale**: Asking for the order, such as "Buy now" or "Order today".
2. **Lead Generation**: Seeking an inquiry, like "Call for details" or "Click here to subscribe."
3. **Third-party Endorsements**: Requesting feedback in the form of testimonials designed to build credibility for your product or service.
4. **Database Marketing**: Cultivating your existing list of customers (past and present) and prospects.

The smart marketer understands that, to be successful, your direct response strategy must incorporate all four levels. Marketing is designed to create interest and the more inquiries and leads you can generate, the more opportunities you have to nurture prospects, which leads to sales. It's an ongoing process of finding leads and converting them into customers.

Switching Channels

There are multiple marketing channels available for your campaign:

- Direct Mail
- Pay-per-click Ads
- Search Engine Optimization (SEO)
- Email
- Live events
- Seminars, Webinars, and Teleseminars
- Social media (blogging, Facebook, Twitter, LinkedIn)
- Newsletters
- Radio Ads
- Television Ads
- Print Ads
- Outdoor Advertising (billboards, bus and subway signs)
- Push Notifications
- Text Messages
- Mobile Apps
- Affiliate Advertising
- In-store Displays
- Press Releases

The best way to increase the success of your advertising is to combine the media that your audience is most likely to access. Your goal is to hit them from various directions with "impressions" (an occasion where your brand is seen). A consumer will rarely act on the first impression, unless you're making an offer that can't be refused! Think about the times you saw an ad or commercial for a restaurant and heard the jingle on the radio. Then you get a post card or email with a "Buy one, get one free" offer and finally decide to head there for dinner. Without all the other impressions collected in your mind, you might not be interested in trying a restaurant you knew nothing about. But the other channels paved the way for you to respond to the special offer.

Consider your target demographic. Then determine at least three channels that you could use to reach them. If you're promoting an app, you're talking to a customer who is comfortable with technology, but that doesn't mean you should limit your marketing to mobile and Internet-based media. Many smartphone users don't text or use Facebook. Quite a few haven't yet grasped the QR code concept. Approach them from other media just to be sure your message will be received.

Going Direct

Let's take a look at direct mail for a moment. This medium is terribly misunderstood by a lot of business owners. They tried it, did it wrong, got poor results, and declared it dead on arrival. If they had just tweaked their message and included a stronger call to action that guided the prospect more clearly, they might have experienced a healthier outcome.

Take a look at the marketing you've done in the past. Conduct a marketing inventory. Itemize everything you did to promote your business—from printing business cards to sponsoring an event to writing a blog. Now think about the message in each one. Did you have a clear call to action? What was the response? Who was the audience?

Take the time to analyze what you've done and you might see a pattern of successes or failures. You might also glimpse an opportunity. What could you have done better? Did you feature a call to action that would resonate with you audience? I've seen people offer 10 percent off their product or service and be disappointed with the weak response. Ask yourself, in the course of your busy life, would 10 percent spark you to change your behavior? Probably not—unless that item had a big price tag where the discount would add up to big money.

Make the offer and the channel fit your customer. The more strategic you are with your direct response campaign—the message and the media—the quicker you will see results.

12 Ways to Use Direct Response

The quickest way to see short-term results in today's economy is through direct mail. The more strategic your direct mail campaign, the more wealth you will create for yourself and the more responses you will get from your requests. Here's your chance to grab market share and dominate!

Consider these 12 objectives that could be established for a direct mail campaign:

1. Generate new customers
2. Capture sales through mail order
3. Identify and qualify prospects
4. Schedule appointments for sales presentations
5. Complement or supplement an existing marketing blitz
6. Test and isolate markets where it's too costly to send out field reps
7. Identify smaller markets that are better suited to direct mail than field sales
8. Broaden the reach of your business (from local to regional, national to global)
9. Communicate with existing clients or prospects
10. Eliminate buyer's remorse by sending customers an after-sale orientation kit

11. Offer a free consultation
12. Build a relationship through targeted monthly communication (e.g., newsletter)

Where Does Your Mobile App Fit?

You might be wondering why I've included a chapter on marketing your business when this book is about a mobile app for your business. Your app is a unique combination of a product, service, and marketing tool. You should promote it with the other channels at your disposal, but also leverage the promotional benefits of the app itself to market your business.

To make your app effective on all levels though, you need to get those downloads! Maximize your success by integrating your app into your current marketing campaigns and across the multiple channels you activate. Once you have the subscribers, you have more customers and prospects on which to build future marketing campaigns to promote your business. It's all tied together!

This is the part where you're looking to see where you can integrate your mobile app into the multiple marketing channels. You're going to have URL links to the App Store, Android Marketplace, and private market stores. You'll have QR codes that can be easily integrated into your direct mail campaigns, on your website, in your social media, on posters at live events, on product packaging, in text message campaigns, on in-store displays, and in print and collateral materials. Remember, your goal is to get your app downloaded, and you need to make it appealing and available to as many people as possible. The larger your list, the greater your opportunity to monetize the app via increased sales and in-app advertising. Learn more here and grab your bonuses while you're at it! www.TheBookOnMobileApps.com.

Connecting the Dots

Success in today's media melting pot of marketing channels requires that you bring your online and offline content together. Once you have accumulated downloads, you have a communication channel to reach out to them in a new way. You can personalize your marketing, which has a 97 percent open rate. What else delivers that type of response?

When you leverage this advantage to create relationships, you're on your way to building a community of interested fans, supporters, and buyers. In this decade, social capital is going to be the determining factor as to which business thrive, survive, or crash and burn. A mobile app is one of the best investments you can make toward accruing that capital. You have the knowledge now to proceed with building that app. You know how to market it. And you know the opportunities that a mobile app presents. All that remains is for you to fire up your engines and make haste to get it done before your competition does. Show your customers that you lead the way, that you speak their language, and that you're not sitting still, complacent with the status quo. Make your move now and reap the rewards of your action.

CHAPTER EIGHT

Post-App Hangover:
3 Steps to Monetize Your App

"Formal education will make you a living. Self-education will make you a fortune."

—*Jim Rohn*

As with any new technology, it's easy to get mesmerized by the shiny newness of it all. But, if you are not using that technology to make money, it's not worth the investment. As I've told you, your mobile app has the potential to generate leads and sales. You can further gain some return on your investment by monetizing it.

If you take nothing else away from this book, set your goals to succeed in three segments—the Before, During, and After. To make money with your app, here are three steps to help you cash in on the mobile explosion.

1. **Before: Establish a Clear Objective.**
   Identify your primary goals for the mobile app you are about to develop. What do you want to achieve with this tool? Are you looking to generate sales and leads, build brand awareness, create social media engagement, or perhaps increase customer loyalty with VIP rewards? Define the results you want from your app and be sure they align with your core business objectives.

2. **During: Create a Strategic Marketing Plan.**
   When your app is ready to go, will you have people anxious to download it? Don't wait until that point to start thinking about marketing it. While developing your app, you need to be developing your marketing strategy. Consider the untapped lead generation potential of app stores—filled with customers who have their credit cards on file and can purchase with one click. How are you going to guide them to your app?

3. **After: Communicate and Sell with a "Push".**
   A mobile app gives you the power of the **push notification.** This amazing feature delivers a **97** percent open rate. And all you have to do is ask your subscriber for permission to let them know about all the great stuff to come from you when they download your app. With their approval, you have the ability to connect with your customers and leads wherever they are, whenever you want (well, okay, within reason—don't be a pest). Give them special offers, information, and tips so they welcome your pushes, and you'll nurture a subscriber into a loyal, raving fan!

Define Your Objectives

There are many reasons you should have a mobile app, as you already know by now. It should be used as a tool to support your business objectives. A mobile app can deliver a broad range of results, but you need to aim it in the right direction. Start with a short list of goals. Then focus on creating the app to meet those objectives.

Above all, don't get so carried away with your pursuit of the Perfect App that you stand in the way of actually finishing and launching it. Yes, there are lots of cool things that your app can do, but keep your eye on your target. That's what you're aiming for. Anything else is just a detour.

Strategize

In the first step, you planned the features and functions you want your app to do. Now it's in the hands of the developer. Use this time to determine how you are going to market your wonderful, new app to the world—or, more specifically, to your target audience.

When your app goes live on the App Store and Android Marketplace, you will get a URL for the direct link. You can then feature the link in your campaign tools. Remember that you'll have to wait until the store has activated your page to get the link. That doesn't mean you have to stall on your marketing plans. Prepare everything so that all you have to do is drop in your new link and print, email, or upload your marketing materials. Use this in-between time to also develop coupon specials within your app that can go live with the launch. You can add and change specials at any time, but be sure that you have something ready for your launch so you'll be prepared to maximize on the added attention you'll get with your "grand opening". This offer also delivers your call to action—"Redeem now!"—so you spark a response and give your subscribers an immediate reward for downloading your app.

You can also create a QR code to connect people with the link on the store of their choosing. Remember that app stores are the new "Google", meaning that people are searching them to find the apps they want and need. With your app on these stores, you are searchable by the site's visitors—conceivably an entirely new list of leads for you. How are you going to harness this wealth of lead potential? What can you do as a marketer to get the attention of your desired subscribers and guide them to the app store where they can grab the download?

The answer is in the "Wow!" factor. Show your prospects how your new app is cool, sexy and fun. You have an audience of app store browsers who are anxious to find something new to explore. Give them a reason to choose you! In addition, you gain the extra "Wow!" here because having an app elevates you to a place as an authority and leader in your particular niche. It builds your brand and communicates your position as an innovator.

Push for Sales

In the aftermath of the launch, you'll activate all of your marketing tools. Be fully prepared to send out the push notifications with your very first subscriber. Tell all of your existing customers about the great benefits of having this app. Your team will be trained to tell customers about the app—keep reminding them to do so, maybe even offer an incentive to them for scoring downloads or handing out cards with the QR code to get the app.

But wait, there is more!

The app store links will be included in your email signature line. Your blog and Facebook page will feature the app with this link. You'll tweet about the launch. Your in-store displays will have a QR code that links passersby to the app store where they can get your app. And they'll do this because your promotional campaign gives them enough reasons to make the download a no-brainer.

You've always got your smartphone close by, but once your app is live, get in the habit of taking your iPad with you everywhere. Use it to show anyone your app on this device, too. Have an iPad at your reception desk and waiting area so visitors can see it. Don't be too shy to share the "Wow!" factor.

When people give you positive feedback about your mobile app, ask for a testimonial. Encourage them to post a review on the app store, their blog, or any other online resource. These reviews contribute to building relationships—both with those people offering the reviews and the others who read them. You will need to amass that growing volume of social capital to guide the success of your app.

Monetize

Now that you have taken the time to set your objectives, build a dynamic mobile app and effective launch your marketing strategy, you now have a beautiful, new list of subscribers. As you maintain your app, be sure to regularly check the analytics to know how many subscribers you have. This is important as you now have leverage for advertising with this information.

The first method for monetizing your app is to begin advertising your products and services within your app as well as to begin a mobile marketing campaign. Banner ads can link to your website to get opt-ins or take them to the App Store where they can download your app. Both options will begin a nurturing process to eventually lead to a sale. These ads can also link directly to a sales page or automated webinar. The ads can link to your Facebook page where you may be running a campaign. Remember, you need multiple sources to fill your sales funnel and marketing within your mobile app will help fill that funnel with multitude of prospects from multiple sources.

Not only can you do In-App banner advertising for your own products within your own app, but as a second revenue source, you can also find other businesses that would be a good strategic alliance to sell advertising space to. Remember, the bigger your list, the more sought after you will be. If you are a Realtor, wouldn't local moving companies just love to get in front of your clients? What if you own a fitness studio and could partner with a local supplement store, nearby yoga studio, and a fitness apparel shop? How about if you have a carpet cleaning business? You could offer local restoration services of another business. If someone wants their carpets cleaned, do they want the windows washed as well? There are a good number of complementary businesses that would and you can easily partner with another local business owner to advertise on your mobile app. If you run a restaurant the serves lunch and dinner, would an alliance with a breakfast and coffee shop make sense for you?

Do you see the potential here? Don't just think outside the box . . . throw the box away! Now that you have this mobile app, how else can you do—and get—business differently? How can you create additional income streams with your new technology that someone else hasn't yet embraced? I bet there are numerous partners out there for you to explore and that your new partnerships will create a win-win for both of you.

The third way to generate more revenue from your app is to add a Shopping Cart feature that links to your PayPal or Merchant account or by using Apple's In-App purchasing feature. If you have products or services to sell, I would highly recommend adding at least one of these for sale within your app. Make it a product or service that doesn't require a long sales process, a product that is easy to supply and easy to service. Apple's In-App purchase program is great because these buyers have a credit card on file and can make one-click purchases. So for instance, if you have a magazine and want to offer a free version with the option of upgrading to the paid subscription, you could easily set up with Apple's In-App purchase program. Be aware though that a cost comes with this ease—a 30% commission to Apple. But getting in front of over 200 million verified credit cardholders can make the commission payment a very wise investment. By simply using an ordinary Shopping Cart, you must guide people through more steps to place an order, which always carries the risk of losing buyers along the way. However, the fees associated with that option only involve PayPal or your Merchant Account.

Keep the momentum

There's a lot of energy around the launch of a mobile app. People are caught up in the newness of it all. The challenge is to maintain that momentum and stay as energized in the months ahead as you were in the first few days. Be vigilant about updating coupons and offers. That doesn't mean you have to lower your prices or give away your profits. That's not the way to make this tool profitable. Be creative about finding new temptations to bring your subscribers back, over and over again. Keep the content fresh and interesting. And never take your subscribers for granted.

Anyone who tells you it's not possible to add revenue quickly with a mobile app is either doing it wrong or has never actually ventured into this medium. Ignore the naysayers. The possibilities are massive! Incorporating a mobile app into your business will dramatically change the way you communicate with your prospective clients. Perch your brand at the top of their minds and do everything within your power to keep it there.

Remember, success is just outside of your comfort zone. Take action and you will see the results. I look forward to seeing you in the App Store. Don't wait—head on over to www. thebookonmobileapps.com to grab your bonuses now!

APPENDIX A: TECHNOPHOBIA? NOT ANY MORE.

Let's clear up the terms!

> *"If we did all the things we are capable of doing, we would literally astound ourselves."*
>
> —*Thomas Edison*

If you're considering getting into mobile marketing and mobile app development for your business—and you definitely *should*—there's a whole other language you need to learn and understand in order to avoid that clueless look when "people in the know" start tossing around terms and acronyms like QR, LBS, push notifications, and MO message. Here's a brief glossary of terms so you can understand this conglomeration of techno-jargon.

4G. This is not four grand or four thousand bucks. The "G" refers to the generation of the mobile technology. 4G is the fourth generation of wireless and broadband wireless data—faster, smarter, and more powerful than the G's that came before.

Analytics. This statistical data tracks the number of mobile app downloads from both Apple and Android app stores, identifying when those downloads took place; used to determine, support, or refine the appropriate marketing strategy in response to the data.

Android. This is a Linux-based operating system for mobile devices (e.g., non-Apple smartphones, tablets). It was developed by Open Handset Alliance, and led by Google, which purchased Android, Inc., the initial developer of this software. Google founded the consortium, Open Handset Alliance, which consists of 86 hardware, software, and telecommunication companies devoted to advancing the open standards for mobile devices.

GPS. The Global Positioning System is a space-based satellite navigation system developed and maintained by the U.S. Government to provide location information any time and anywhere on Earth that can be seen by one of the 24 satellites orbiting the planet.

GPS Coupon. These location-based coupons require the user to "check in" at a location designated by the GPS Coupon; the offer can only be unlocked at the specified location.

GPS Direction. Turn-by-turn directions provided to a mobile phone user via the native GPS navigation feature on the phone.

Hosting. For mobile apps, hosting is a service that provides an interface or control panel for managing the mobile application. It also provides third-party support for push notifications and software upgrades and changes.

HTML5. HyperText Markup Language (HTML) is the primary language used for the structure and scripting of web pages. HTML5 is the web-based version of a mobile app. HTML5 apps are also called Web Stack Apps (e.g., BlackBerry). If you do not have a mobile device that connects to the Apple or Android store, you can still run an app with HTML5 technology. Look for an independent app store to open, just for HTML5 apps.

iOS. Formerly known as the iPhone OS, the term for Apple's mobile operating system has been shortened to iOS, which now supports not just the iPhone but the iPad, iPod Touch, and Apple TV.

iPhone. Apple's line of Internet and multimedia-enabled smartphones. Initially closed to third party developers, Apple opened up its iOS (Internet Operating System) to developers in order to expand the library of Apple-approved software applications for this device, which currently has more than 500,000 apps in the Apple App Store.

Location-Based Services. Using GPS locators, a mobile app can determine a mobile device user's location and customize geographic-specific advertising messages.

M-Commerce. The buying and selling of products and services using a mobile device.

MMS. Multi-media messaging service goes beyond just text messaging—also known as SMS or Short Message Service—and include graphics, audio, and video.

Mobile App. Those cute, square icons on your smartphone represent a software app—short for "application"—that is specifically designed to run on a mobile device (as opposed to a desktop or laptop computer with a larger screen).

MO Message. A mobile-originated message is a text message that is sent from a mobile device.

Newsstand App. A feature within Apple's iOS5 mobile operating system, the Newsstand app provides a coherent means for purchasing magazine and newspaper subscriptions. The app's icon is unique in that it is not the typical square icon, but a customizable cover aligned to the latest issue on the newsstand. An app within the Newsstand umbrella will automatically download new issues through daily push notifications.

One-Touch Calling. This feature enables a mobile phone user to dial a business without entering the phone number.

Push Notification. Apple and Android provide this service, which uses "push technology" to send data through a constantly open IP connection (server) to forward notification via badge, pop-up, or alert from the servers of third-party applications to the mobile device. Mobile apps don't need to be open or running in order for the recipient to see the message. Instead, the push notification passes raw message data straight to the application, which has full control of how to handle it. The application might port a notification or silently sync data without the user being aware of the action. A push notification could be a text message alert of a special offer, news update, or other item related to the app's function.

QR Code. The Quick Response Code is a two-dimensional matrix barcode that is readable by smartphones equipped with cameras and a QR reader software program. By scanning the QR Code, the user is automatically redirected to a Web page that provides more information and/or offers specific to the product with the code—a function that has increased the code's appeal for marketing to mobile users. QR codes were created in 1994 by Toyota subsidiary, Denso Wave, and designed to be decoded at a high rate of speed. They were first used to track vehicles during the manufacturing process.

QR Coupon. Utilizing the QR Code's two-dimensional pattern, coupons link a mobile user to a particular discount by requiring the user to scan a QR code to unlock the offer.

QR Scanner. The scanner uses the device's native camera feature in conjunction with QR reader software to capture a picture of a QR code. The software decoder then transforms the data that is encoded into the symbol and translates to an action, such as redirecting to a URL. The scanning occurs in seconds, performing an instantaneous transaction from the user's mobile phone to the mobile web.

Shopping Cart. The shopping cart feature in an app enables consumers to engage in electronic commerce (e-commerce), the buying and selling of products or services over electronic systems, such as the Internet. E-commerce utilizes a virtual shopping cart in which consumers add products or services to be purchased, prior to checkout.

ABOUT THE AUTHOR

Cheryl Paulsen spent eight years serving in the United States military (Navy and Reserves), including a recall to duty after September 11. She served on the guided missile cruiser USS Port Royal and, after completing Nuclear Power School, Cheryl was stationed aboard the aircraft carrier, USS Theodore Roosevelt.

She combined her powerful interest in training and an unwaverable passion to help small business owners grow their business with new technology. Using her experience, Cheryl founded Paulsen Enterprises and Mobile Apps for Small Business, a company that focuses on developing mobile applications for small businesses and entrepreneurs.

Cheryl is also an impassioned speaker. Her topics include Technology for Small Business Owners, Women and Leadership, and Profitability Leveraging Today's Mobile Technology.

In her spare time, Cheryl and her husband raise oysters for the Chesapeake Bay Foundation. Because every adult oyster filters 50 gallons of water per day, increasing the oyster population on the reefs are helping to clean up the Bay.

Cheryl holds a Bachelor's degree in Engineering Chemistry from Rensselaer Polytechnic Institute and a Master's in Engineering Management from Old Dominion University. She has a Certificate in Naval Nuclear Power Training from the Naval Nuclear Power School.

For more information, contact Cheryl
www.MobileAppsForSmallBusiness.com.

CLAIM YOUR FREE GIFTS NOW!

MORE BONUSES AVAILBLE AT www.TheBookOnMobileApps.com

BECOME A SMALL BUSINESS RENEGADE™—FOR FREE!

($407.97 Worth of Pure Money Making Information)

Cheryl Paulsen is offering an incredible opportunity for you to see _why_ Small Business Renegades™ is _**the place**_ for small businesses that are seeking to navigate the treacherous waters of successfully marketing your business. Get $407.97 worth of pure _**money-making**_ information—including a _free_ month as an 'Elite' Gold Inner Circle Member of Small Business Renegades™.

Here's what's included with this free offer:

- **Small Business Renegade University: Webinars** (Value: $387.00)

  Learn how to change the game and fuel the app addiction so you can PUSH profits to put more $$$ in your pocket.

  - How any business can multiply income by 10X

  - The importance of a mobile app in your sales process

  - How to use the latest technology to instantly reach your clients

  - 3 most important steps to monetizing your mobile app

- **'Elite' Gold Inner Circle Membership** (Value: $19.97)

  An issue of the Small Business Renegade Marketing Letter. Each issue overflows with the latest marketing and mobile strategies, plus terrific examples of "What's Working Now" strategies, timely marketing news, trends, and ongoing teachings. As soon as it arrives, you'll want to find a quiet place, grab a highlighter, and devour every word!

- **Exclusive "Members Only" Perks:**
  - Special member teleconferences every month
  - Restricted access website
  - Continually updated *Million Dollar Resource Directory* with valuable contacts and resources used by Cheryl and her clients.

To activate your ***most incredible free gift ever***, you pay only a one-time charge of $3.95 ($6.95 for international subscribers) to cover shipping.

After your one-month free test drive, you will automatically continue at the lowest price of $19.95 per month. Your credit card will not be charged for the monthly fee until the beginning of the month after your free trial. You get a full month to test and profit from all the powerful techniques and strategies that will change your mobile game. It's impossible for you to lose, because if you don't absolutely *love* everything you get, just cancel your membership before the next month and never get billed a single penny!

You can cancel your membership at any time by calling Mobile Apps for Small Business at 410-827-6236 or faxing a cancellation note to 888-378-1118 (Monday-Friday, 9 am – 5 pm).

Name _____

Business name _____

Address _____

City _____ State _____

Postal code _____ Country _____

Email* _____

Phone _____ Fax _____

Credit Card Instructions to cover $3.95 ($6.95 Int'l) for shipping:

_____ Visa _____ Master Card

_____ American Express _____ Discover

Name on card

Credit card number _____ Exp date _____

Signature _____ Date _____

*EMAIL IS REQUIRED TO NOTIFY YOU ABOUT THE SMALL BUSINESS RENEGADES WEBINARS

Providing this information constitutes your permission for
Mobile Apps For Small Business to contact you regarding
related information via, mail, email, fax, and phone.

FAX BACK TO 888-378-1118
Or mail to: 600-B Abruzzi Drive #125, Chester, MD 21619